Surgical Back Fusion
The Decision, Preparation, and Recovery

Martha Gilliland lives in Tucson, Arizona. She is a retired professor of geology and environmental engineering, a former Vice-President of the University of Arizona, Provost of Tulane University and Chancellor of the University Missouri-Kansas City. She is the proud mother of two and grandmother of six wonderful people.

Table of Contents

i. **Prologue:** Three Steps for Success 1
ii. **The Decision:** How do You Decide Whether or Not to Choose Surgery? 3
   1. First, do not rush 3
   2. Second, check your attitude 3
   3. Third, choose a surgeon 4
   4. Fourth, take time to understand the procedure 5
   5. Fifth, be clear with yourself what you want after the surgery and whether or not the surgery can produce that 7
   6. Sixth, understand the risks 7
   7. Finally, be realistic about what is possible 8
iii. **Preparation:** How to Get Ready For Surgery 11
   1. First, prepare physically 11
   2. Prepare mentally 12
   3. Obtain the gadgets you will need 13
   4. Finally, make your plan and include friends and family in it .. 13
iv. **Recovery:** The Short Term, The Long Term, and The Unexpected 17
   1. The first six weeks 17
   2. The second six weeks 17
   3. The unexpected 18

Three Steps for Success

What do you do when your physician says that you need back surgery, in my case a fusion of the two vertebrae at the bottom of my back? Should you do it? When should you do it? How do you decide? How do you prepare?

I faced these questions in 2019. My back hurt. I had shooting pain down my left leg and sometimes my right leg. My left foot sometimes tingled, even became numb. I had experienced some of these issues for ten years, but now it was bad. I had fallen twice while hiking. I needed to do something. Various surgeons told me that surgery could fix the problem. But I was skeptical. I had heard mostly negative stories about the success of back surgery. And I have a bias against surgery, generally believing I can fix whatever the problem is with a less invasive procedure. Yet, the prospect of living with what I had no longer seemed like a viable option.

In 2019 I was 74 years old. I have always been an active person, accustomed to hiking, bicycling, and golfing extensively with friends, children, and grandchildren. I wanted to be able to continue these activities. My back pain was interfering with all of them. I had already tried all kinds of therapies, including massage, Pilates, yoga, acupuncture, chiropractic adjustments, and cortisone injections. Each one helped for a short time, some for several years, but eventually nothing worked. Now I was at a crossroads; I had to decide – live with what I have or take the risk of surgery.

Although I am a scientist, I know very little about human anatomy or medicine. I am the type of person, however, who asks lots of questions and digs deeply into the information on the internet. Consequently, I became obsessed about understanding the anatomy of my lower spine, what was causing my pain, and what the surgeon was going to do to fix it.

This is my story and the lessons learned as I made the decision to proceed with a back fusion of my lower back.[1] In retrospect I know I would have benefited greatly from easier access to this information. As I asked questions, I quickly discovered answers to questions about risks and recovery are fragmented, sometimes inconsistent. I spoke to many types of health providers in advance of making a decision. The nurses and physician's assistants are a wealth of information, but they do not volunteer it unless asked. Most of us do not even know what we need to know, what questions to ask, or who to ask. Answers to one question leads to more questions and more struggles for answers. It is not surprising that many people, me included, find the process frustrating and confusing.

Three steps for success: (I) The Decision: what do you need to know to make the decision to proceed with surgery or not? (II) Preparation: once decided, how best do you prepare? (III) Recovery: after surgery, what should you expect during recovery in the short and long term? Some of this information is based on my own research; some of it suggests where and how to find answers for yourself; and some simply relates my experience. Pick and choose the parts of that fit your situation.

---

[1] Bone fusion two-levels -- L4 to L5 and L5 to sacrum. The L stands for lumbar vertebra; L4 and L5 are at the bottom of the spine. In my case, the discs that separate these vertebrae had degenerated and flattened out, so that the space between vertebrae was compressed and the vertebrae pressed on the big nerves that come out of the spinal cord and run across the buttocks and down the legs to the toes. A common version of this is sciatica. I kept that spacing open just enough for many years through core strengthening exercises. But eventually my legs did not work correctly. The fix is a laminectomy with fusion. First, they cut out the lamina which is the part of the vertebra that forms the arch in the back. They save that bone for use later in the surgery. This allows them to pull the vertebrae apart and open up spacing. They then put titanium screws and rods into each vertebra to make it strong and stable (needed because the lamina is gone). They take a piece of my lamina bone and build a bone wall between each vertebra with a titanium band around it. That's it. It's done. This keeps the vertebrae apart (decompressed). That bone has to fuse and that takes 6 weeks to get really going and will continue for months. Our bodies know how to heal broken bones really well. The body reacts in a fusion just like it would if you broke your leg. It makes osteoblasts and deposits clusters of them to make a bone.

The Decision:
How do You Decide Whether or Not to Choose Surgery?

First, do not rush.
Rarely is back surgery an emergency. Back pain is extremely common. Surgery will not necessarily fix it. Many different problems can cause back pain, and pain can often be alleviated without surgery. For example, strengthening the abdominal muscles that hold up your spine eliminates many types of back pain. Physical therapists are experts in designing such a strengthening program.

Ask lots of questions. Become fully informed about what is causing your pain and what the recommended solution is, so you can make the decision that is right for you. Nearly all of the literature recommends that surgery be a last resort. Deciding if and when to have surgery is more of an art than a science. You can be sure that someone else's experience will not be your experience. Everyone is different. Take your time.

Second, check your attitude.
I was having lunch one day with a friend, who is also a physician. I told her I was considering a back fusion. I shared that I had quite a negative view of any surgery, especially back surgery. I indicated that I felt like a failure because I had not been able to fix the problem with proper exercise. I expected my friend to agree with me, but she said, "Don't proceed with surgery with that attitude. Negative attitudes can produce negative outcomes." This stopped me dead in my tracks!

Then I learned that my friend's observation was supported by research which suggests that our attitude about health does impact the outcome positively or negatively.[2]

Feeling vulnerable, I asked friends and family to help me shift my attitude. They each had much the same response: "Martha, you are always going on adventures, bicycling or climbing some mountain. Just rename this an adventure and figure out to make if fun." They made suggestions on a plan for the recovery that included something new and interesting every day, even ideas on what to "binge watch" on Netflix. They helped identify adventures I could do in the future once this problem was fixed. With that plan, I finally was able to name the surgery a health adventure. I pictured primarily opportunities rather than problems. Of course, if your attitude about surgery is already positive, keep it that way.

Third, choose a surgeon.
Choosing a surgeon is the most important decision you will make in this process. This is not like choosing someone to fix your car or the roof of your house. If your car mechanic or roofer turns out to be incompetent or makes a mistake, you get a second chance. With back surgery, there are no "do-overs." This is your body and your future. In finding the right surgeon,
- Ask your primary care physician for a recommendation.
- Talk with friends who might share useful information or at least help you ask the right questions.

---
[2] *Your Medical Mind: How to Decide What is Right for You* by Jerome Groopman and Pamela Harztband

- Be sure the surgeon you pick has performed the surgery hundreds of times. Sometimes the website for a surgeon provides that information, and usually competent surgeons are more than comfortable answering the question. If not, your primary care physician can likely find the answer.
- Find a surgeon who understands what you want in your quality of life, respects what you want, and will listen to you. I was fortunate to find a surgeon who cared about hiking and bicycling. Eventually, I picked a two-person surgical team, both of whom are big hikers. One would perform the first part of the surgery and the other the remaining part. Both doctors would be present throughout the entire surgery.
- Compare your own biases about surgery with those of the surgeon. My primary care physician recommended that I read *Your Medical Mind.* The book helped me clarify my own biases and ask good questions of my surgeons. Most surgeons are highly competent people, but they have a perspective about surgery that is likely different from mine or yours. Find out how different. No bias is right or wrong.

Fourth, take time to understand the procedure.
What exactly is causing your pain and what is the surgeon going to do to fix it?

I had an MRI to diagnose the problem. It read: "severe L5-SI degenerative disc disease, spondylosis, and anterolisthesis." I had no idea what these terms meant. When I asked the surgeon to demonstrate their definition using the model of a spine sitting on his desk, I could clearly see that the disc at the bottom of my spine was nearly flat.

Consequently, the vertebra above it was compressing the nerve coming out of my spine and running down my leg. There was just no space for the nerve.

It was pinched, and so the vertebra kept banging on it. And one vertebra was also a little crooked. No wonder I had shooting pain. With the same model, the surgeon showed me how he would pull apart the vertebra, creating space for the nerve but then would have to screw that vertebra tightly to the one above it to prevent collapsing. That procedure is the fusion. I watched several *YouTube* videos of the procedure. The bottom of my spine would no longer bend. The fusion would stop the pain but being "fused" sounded terrible, like something out of a *Wile E. Coyote* cartoon.

I had worked hard during the years previous to 2019 to strengthen my abdominal and back muscles so the muscles were strong enough to hold open the spaces for the nerves. But strengthening exercises were no longer helping; the degeneration was too severe.

Armed with this understanding, try to obtain a second opinion about the procedure. Simply obtain copies of your X-Ray and MRI images and ask a second surgeon to provide an opinion. The images and reports are easily obtained by signing a form at the image center. The images are provided on a computer disc or as email attachments. I submitted them to Barrow's Neurological Institute for a second opinion.[3] Barrows offers access through their website. By submitting images and paying a manageable fee, a specialist at Barrows analyzes the images and provides an opinion on how best to proceed. In my case, the opinion aligned with the recommendation from my surgical team. That was comforting.

---

3 https://www.barrowneuro.org/patients-families/get-a-second-opinion/
Get a Neurological Second Opinion, Barrow Neurological Institute

Fifth, be clear with yourself what you want after the surgery and whether or not the surgery can produce that.
I wanted to be able to hike and bicycle without constantly worrying that pain would shoot down my left leg causing me to fall. I wanted to be able to turn over in bed without pain shooting down my leg. Hiking and bicycling are essential to my quality of life, because I like being outside and because many of my friendships are built around these activities. You may not be as interested in intense activity but get clear about what is important to you. I asked my surgeon, "Will I be able to hike, bicycle, and golf and how long will it take for me to return to these activities after the surgery?" My surgeon said, "Nothing is certain. Restrictions will be significant for six-weeks, but you will be walking immediately. You are physically fit, and so I expect you to do quite well. You will likely be able to resume all activity after three months. Full healing will likely require six to nine months with gradually increasing activities permitted along the way. And, of course, there are risks."

Sixth, understand the risks.
No surgery is without risks, but I wanted more than my surgeon's assessment of those. I read articles on various websites [4] about the risks of a surgical back fusion and quickly learned that the biggest risk was damage to a nerve during surgery, which could cause paralysis or loss of bladder control. These, however, are said to be quite rare, but are higher if you are a smoker, diabetic, or obese. I was none of those.

---

[4] https://healthfully.com/214994-complications-of-lumbar-fusion-surgery.html
Complications of Lumbar Fusion Surgery by Catherine Schaffer, *Healthfully*, Aug. 2019.

Occasionally, the titanium screws or rods that hold things together after surgery come loose before the bones fuse. Refraining from falling would minimize this risk over the first months.

But, as I dug deeply into the plethora of articles on medical websites, I learned that it is not uncommon for pain to persist in the hips, not shooting pain but pain. Just when I was about to say yes to surgery, I learned that, at my age, ongoing pain is not uncommon. This caused me to send a list of questions about risks to one of my surgeons. He called a few days later and patiently answered each question. His patience and knowledge gave me confidence that I had chosen the right surgeon. But many of his answers were still "maybe or likely or probably." There simply was no certainty about the outcome. It is a complicated process.

Pain medication also poses risks. Opioids are prescribed after most surgeries. I knew of people who had become addicted after relatively simple hip and knee surgery. It would be important to be vigilant about weaning myself off of those opioids. Additionally, anesthesia has side effects, usually producing confusion, dizziness and nausea during the first day or so. But side effects can also occur over the long term. I had some temporary hair loss, which is not uncommon but certainly disturbing.

Finally, be realistic about what is possible.
During a discussion of risks, my friend said, "Martha, you need to set realistic expectations." That startled me. I realized that, generally, all of us feel better when the outcomes of any challenge, whether a test, a race or surgery, exceed our expectations rather than disappoint us.

Disappointment leads to frustration, sometimes depression, which can inhibit healing. I tend to set high expectations for most of what I do in my life. Sometimes this is a beneficial characteristic since it causes me to work hard and stretch my mind or body in order to achieve my goals.

I have learned, however, that my high expectations can be unrealistic. As it turned out, *you are not always as young as you feel* -- I did not fully appreciate that I was 74 years old and was not likely to recover as well as a 54 year old person. Consequently, I had some disappointments during the recovery process. You almost certainly will say to yourself, "why didn't they tell me this in advance or why didn't I think of that." Unexpected outcomes occur. Just know that before you make your decision to proceed.

Summary
Make a mental picture of exactly what outcomes from the surgery you desire, while at the same time understanding that your ideal outcome may not be possible. But find out if the outcomes you want are probable. What physical activity do you want? What routines do you want in your life? Will these be possible after the recovery? What can go wrong? Take your time finding a surgeon and be sure you like him or her. Keep asking questions until you gain clarity. Then, as you sit quietly reflecting on what you have learned, make the decision and set the date. Ultimately, when I knew I could not go on with the pain I had, I chose to take the risk and proceed with the surgery. But I still left myself four months in advance to prepare. Do not rush. Being prepared is essential for success.

Location of Lumbar Vertebrae

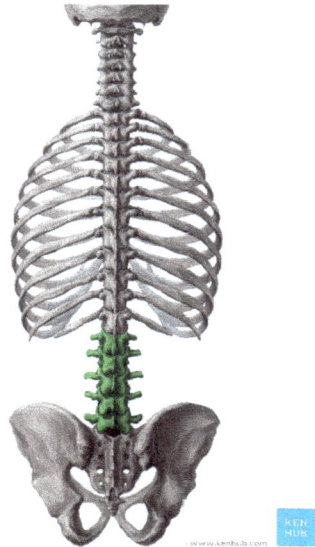

Alice Ferng B.S., MD. "Lumbar Vertebrae." *Kenhub*, Kenhub, 8 July 2020, www.kenhub.com/en/library/anatomy/lumbar-vertebrae.

Preparation
How To Get Ready For Surgery

Lack of preparation is a major cause of problems after surgery. The healthcare system is not designed to help you prepare. You must take charge of this and allow time for it.

First, prepare physically.
Make sure your surgeon prescribes physical therapy in advance of surgery. You need to have strong muscles in the right places. The physical therapists can diagnose where you are weak and prescribe exercises for strengthening.

I also discovered that the physician's assistant in the surgeon's office is a treasure trove of information on how to prepare physically. She is the one who fields the questions from people after surgery, and so she understands what problems people have after surgery and which could have been avoided with proper preparation. Spend some time with her asking questions. I asked, "what is the worst thing that could happen after surgery." She said in the strongest of tones, "Martha, you cannot fall; a fall during the first six weeks will put you right back on the operating table."

She also introduced me to the no "BLT" rule – no bending, lifting, or twisting during the first six weeks. She said I would need a special way of standing up since I had to do so without bending forward. The trick, she said, is to look at the ceiling, which will keep you from bending. This particular move uses different muscles in your legs than the normal way of standing up. When I first tried this prior to surgery, I was surprised by its difficulty. I practiced it regularly before surgery, to the point where it was easy.

Also, I learned about the "log roll" as the process for getting in and out of bed, and I practiced that. I asked, "What about going up and down steps indicating I have a two-story house." To my surprise, I learned this would not be difficult since I was accustomed to going up and down steps anyway. It is much easier to keep your back straight climbing steps than it is standing up from a chair or toilet seat. She told me I would not be able to drive a car for four to six weeks. Major pain medications would be needed, including opioids. I would be up and walking a few steps the afternoon after surgery and would need to take short walks often during the day after I was home. I would need help at home for at least a week.

Had I not been persistent with my questions of the physician's assistant, I would not have known these "rules" until discharge from the hospital. Consequently, I would not have been prepared for recovery. Take the initiative and understand how to prepare for surgery and set up for recovery. If you do nothing else, work on your strengthening the right muscles.

Prepare mentally.
Four components define being prepared mentally. First, be sure your attitude is positive. Second, have a plan in place for recovery. Third, be at peace in case things go seriously awry and you never wake up. And fourth, do guided imagery. Research at the Cleveland Clinic[5] demonstrates that patients who listen to guided imagery tapes prior to and after surgery rated their pain lower and required less pain medication than the control group. Numerous podcasts now provide guided imagery sessions. Simply google "guided imagery podcasts for pain," and you will find many.

---
5 https://my.clevelandclinic.org/health/articles/11307-pain-control-after-surgery
Pain Control After Surgery, Cleveland Clinic

Obtain the gadgets you will need.
For me gadgets included a walker, shower stool, toilet riser, and several grabbers. Grabbers are these nifty devices which help you pick up items from the floor without bending. They also are essential for dressing yourself without bending. The hospital will likely provide the walker and one grabber. I purchased the other items through Amazon, and I purchased extra grabbers so I could position them in several rooms. I found my grabbers to be really wonderful. That little device can pick up almost anything, from a food scrap, paper, socks, shoes, clothes, towel and even an iPhone. Using the grabber, I was even able to load my washer and dryer, taking one piece of clothes at a time, picking it up with the grabber and tossing it in. I still use it to reach into high places or pick up something under the bed or behind the washing machine.

Finally, make your plan and include friends and family in it.
I knew I would need significant support during the first six weeks and probably longer. Like many of us, I am reticent to ask for help even though cognitively I know people want to help, and I know that, if the situation were reversed, I would want to help. But part of my attitude shift was making a plan that felt like it might be joyful.

Ask people who you love to be with you. My approach may not be appropriate for you; just make sure you have people around you who you trust and love. I asked my two adult children to be with me during the initial week. They do not live in Tucson and are busy with their families and demanding jobs. Traveling to Tucson would require a plane ride and time away. Therefore, I did not want to put them on the spot by asking them on a phone call.

I wrote them a letter, provided flexibility in dates, and asked one or the other to be here during the first week. My son came for the first three days of the hospital stay. My daughter came on day three and stayed for four more days. They overlapped in my hospital room for three hours. The time I spent with each of them was spectacular, because I so rarely see them by themselves. While I dearly love their spouses and children, sharing time with each of them one-on-one was a positive aspect of the surgery. Before surgery I had fun buying their favorite foods. My son likes a specific type of Greek yogurt for breakfast. My daughter likes protein shakes. We shared the types of conversations about life that are only possible with one's children.

    My son and I watched the US Open Tennis tournament in the hospital room while munching on his favorite type of cheese and crackers. He brought me my favorite coffee. He advocated for me when it was time for pain medicine, knowing the nurses are always quite busy. He also pointed out that I should not be writing emails, because I was a little confused and lacking in good judgement from the effects of the anesthesia. He was right.

    My daughter brought me home and got me settled. She set up a chart delineating what time of day I was supposed to take which of the four types of medicine and when I should start cutting back on the opioids. I generally consider myself quite organized, but I needed that chart in order to remember when to take which pill. She reminded me that my only job for six weeks was to heal: "just sleep, eat the right foods, take the medicine at the right time, and take short and very frequent walks." She also has an "eye" for interior design, and so she suggested some changes to my house. She was right.

She was a major comfort during the day and night and her competency eliminated my worries. We shared wonderful meals together – just the two of us. If you do not have a spouse or adult children who can help, reach out to friends or hire an advocate through a care giver service for a few days.

I asked four different friends to each spend one night in my guest room at home for the first four nights after my daughter left. Their phones were on alert during the night. Most nights, I woke with intense pain and was able phone my friend in the guest room to bring an ice pack. These friends handled meals, managed my pain medications, and stayed close by while I showered. We had some fun talking, although I think I mostly slept. If you are not in a position to have friends or family spend the night, the surgeon will send you to a rehabilitation center, which is often paid for through your health insurance.

For the next five weeks, I planned visits from friends for breakfast and late lunch. One of my friends sent a sign-up sheet to the list of friends which I provided her. She listed time slots for people to come to my house, and she included very specific directions: *bring or make breakfast at 7:30 AM or bring or make lunch at 1:30 PM; don't stay more than 90 minutes. Be prepared to take a short and slow walk. And please note that she does not eat red meat.* Five weeks of time slots were filled in 3 days. Wow; that felt nice. Relaxed time to visit with friends is rare, and so this also was a special outcome.

I also asked my friends to load and unload my dishwasher, reach into the bottom of my refrigerator to retrieve items that would require bending, pick up a few things at the grocery store, and change the sheets on my bed.

I could *swifter* my floor but not sweep up the little pile of debris that it generated. My friends did that. I had, in advance, bought a small gift for each helper, handing it to them as I said, "thank you."

Even if you do not live alone, your partner will want some breaks from waiting on you. Intersperse some friends during those first six weeks to give you both a break from each other.

In short, my children and my friends were a special gift. Spending time with them was a benefit of having the surgery. I treasure that time.

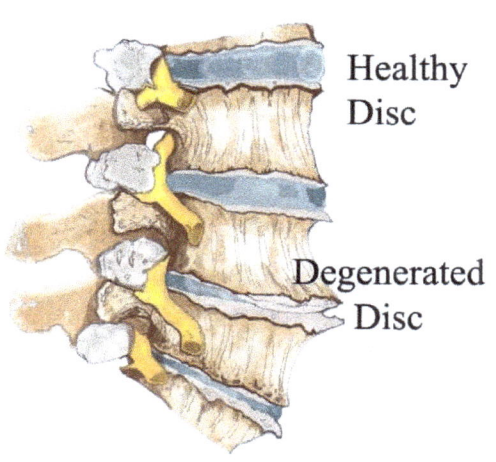

Seto, Joe. "Rupert Health Centre Inc." *Rupert Health Centre Inc Blog Lower Back Pain Degenerative Disc Disease Comments*, www.ruperthealth.com/blog/uncategorized/lower-back-pain-degenerative-disc-disease/.

Recovery
The Short Term, The Longterm, and The Unexpected

I am now one-year post-surgery. I am happy I did it, but recovery was harder and longer than I expected.

The first six weeks

The First six weeks went exactly as I expected. My plan worked. Thanks to the preparation, I felt strong; I slept well; I had great conversations and short walks with friends; I read and watched television. Friends brought food. My gadgets worked well, although I did not really need the walker or shower stool for more than a couple of days. My fitness preparation worked well. After the first three weeks, I did struggle with my patience. I wanted to be active. I wanted to drive. Each time I had some pain, I worried something had gone wrong. The words of the physician's assistant loomed large, "Martha do not fall."

Finally, it was time for the six-week X-Rays and review by the surgeon. I remember the relief when he said, "All looks great. I will see you next in three months. Start physical therapy; you will be stiff and will have lost some strength. But, do what the therapist tells you to do, and you should be fine."

The second six weeks.

As I started physical therapy, I was impressed by the knowledge of the physical therapist about the exercises required to re-engage the muscles, but I was quite surprised by the stiffness in my hips and legs. I followed her instructions each day.

Serious commitment to physical therapy is absolutely essential, and I benefited greatly from it. That said, I was discouraged at times by how slowly my body seemed to respond and how much muscle strength I had lost just during that first six weeks.

The unexpected.
About three months after surgery, as I became more active, I developed unexpected pain on my right side -- from the top of my hip, extending around the side into my groin and then down my leg, sometimes all the way to my ankle. I could bicycle, but I could not walk for more than 30 minutes without setting off intense burning pain down my right side. This pain also often woke me up at night, causing me to go in search of ice packs to settle it down. The pain was quite bad at times and very discouraging.

My surgeon said, "I believe you have hip girdle and hip joint pain; I think it will resolve over time. Everyone has asymmetries between the left and right sides of their bodies. Over the years, each person learns to walk and use muscles in a manner that compensate for the asymmetry. The fusion undoubtedly fixed some of the asymmetry. Now you have to readjust. And, truthfully at your age, you also have some degeneration and arthritis in there." This made sense to me, but I had not expected it. Worse, I did not know how to "fix" it.

While it seems obvious to me now, a fusion normally causes an individual to walk differently. I had been walking for many years in a way that minimized my back pain. Now I have a different walking gait using muscles differently, which in turn, require different joint movement.

This caused new pain as these muscles stretched and strengthened and the joints adjusted. This was disapointing and unexpected.

At home, I googled hip girdle and was bombarded by a lengthy list of girdles for women, beginning with a sale at Walmart. Information under "hip girdle pain" was more helpful, but I still found it difficult to understand. As I read, I was reminded of the children's song, *Dry Bones*. The song teaches kids body parts. The kids sing and dance as they touch each body part, *The back bone is connected to the hip bone; the hip bone is connected to the leg bone* and so on.

It turns out that the part of the song about the hip bone being connected to the leg bone omits a great deal of detail. To my surprise, that hip girdle has five bones itself, four of which I had never heard of – the sacrum, ilium, pubis, ischium, and pubic symphysis. These bones and associated ligaments and joints are connected to form a ring. The legs connect into that ring at the hip joint, and that ring and joint handles asymmetries. My asymmetries were fixed in four hours in surgery. At 74 years old, this ring was just not able to adjust quickly, and I had arthritis. The fusion changed my posture for the better. The old muscle imbalance was gone, but my joints, ligaments and all they connect into were having trouble adjusting.

I followed the advice of physical therapists and struggled to be patient. Most of the time, I could motivate myself to do the work, but it was hard. I returned to my Pilates class, which was and still is extremely helpful. I did the stretches recommended by the physical therapist in the morning and evening. Bit by bit over six months, the pain subsided. Now at one year after surgery I can hike without pain, no matter the distance.

I can bicycle and I can golf. I still occasionally have burning pain on my right side.

This is where those realistic expectations come into play. Fixing a body part with surgery will not make you into a 20 year-old. I am happy that I no longer have shooting pain down my leg, but my body, at 75 years old, still has stiff places and places that sometimes hurt.

No one except you will know if and when it is time for surgery. Recovery from a back fusion will require substantial commitment and time. Only you can weigh the possibility of getting rid of the pain you have now against the risks of having surgery combined with the recovery effort required after surgery. When I finally proceeded, I knew recovery would be hard, and it was a lot harder than I expected. But now it is all behind me and it was worth the effort.

My one piece of advice.
Take your time in making your decision, ask lots of questions and do the work to prepare. There are no "do-overs."

Titanium Screws and Rods fusing the vertebrae in my back.

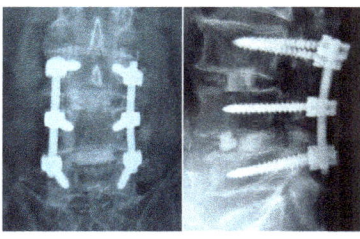

Boulder Neurologic Institute, Boulder, CO

www.ingramcontent.com/pod-product-compliance
Lightning Source LLC
Chambersburg PA
CBHW041920240526
45473CB00038B/2915